Education 250

以爱的名义

In the Name of Love

Gunter Pauli

[比]冈特·鲍利 著

[哥伦]凯瑟琳娜·巴赫 绘

何家振 译

上海遠東出版社

特别感谢以下热心人士对童书工作的支持：

匡志强　方　芳　宋小华　解　东　厉　云　李　婧

刘　丹　熊彩虹　罗淑怡　旷　婉　杨　荣　刘学振

何圣霖　王必斗　潘林平　熊志强　廖清州　谭燕宁

王　征　白　纯　张林霞　寿颖慧　罗　佳　傅　俊

胡海朋　白永喆　韦小宏　李　杰　欧　亮

目录

Contents

海蛇和虎鲨正在观察河豚在海底的沙地上画美妙的图案。海蛇评论道：

“我打赌你喜欢吃那条河豚。”

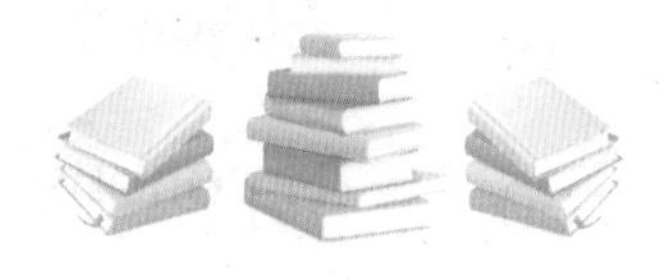

A sea snake and a tiger shark are observing a pufferfish making marvellous designs in the sand on the ocean floor. The sea snake comments:

“I bet you would like to eat that pufferfish.”

海蛇和虎鲨……

A sea snake and a tiger shark ...

河豚味道很好，但是有毒……

Pufferfish are very tasty, but poisonous ...

“噢，我是喜欢吃。但并非只有你我喜欢吃河豚。人类也认为河豚是美味佳肴。”

“河豚味道很好，但是有毒。”

“我想你我都是少数能吞下河豚而不受其致命毒素影响的幸运儿。”

“Oh, I do. But you and I are not the only ones that like to eat them. People consider pufferfish a great delicacy.”

“Pufferfish are very tasty, but poisonous when eaten.”

“I think you and I are amongst the happy few who can devour a pufferfish without being affected by his deadly poison.”

“真的，我们能把河豚连同毒素一起吃掉，不会有任何麻烦。但是，你知道一条河豚的毒素足以杀死30个人，人类应该非常小心。”

“令人吃惊！这些河豚看起来像金鱼一样温顺。它们非常缓慢地游动，却是致命的。”

“Indeed, we are able to eat them, poison and all, without any trouble. But knowing that one fish contains enough poison to kill thirty people, people should be very careful...”

“That's surprising! These pufferfish look as docile as goldfish. They move around ever so slowly, yet are lethal.”

看起来像金鱼一样温顺，却是致命的

Look as docile as goldfish, yet are lethal

就会卡在捕食者的喉咙里……

It will stick in its throat ...

“饥饿的捕食者以为河豚很容易抓到，但当他的嘴被一团带刺的毒物填满而停下来时，那他就不认为自己是幸运儿了。河豚膨胀起来，就会卡在捕食者的喉咙里。”

“又有谁会想到，这种笨重的鱼竟然会是水底世界最伟大的艺术家之一呢？”

“A hungry predator thinking a pufferfish an easy catch, may no longer think himself so lucky when he ends up with a stinging ball of venom in his mouth. As the pufferfish puffs up, it will stick in its throat.”

“And who would have thought that this bulky little fish would turn out to be one of the greatest artists of the underwater world?”

“是啊，我以前见过雄河豚做沙雕。很少有人意识到他们是多么伟大的艺术家。”

“他们拍动着鳍，不知疲倦地日夜劳作，创作出如此完美的几何艺术作品！”

“你知道小雄河豚是用他压碎的贝壳碎片装饰雕塑艺术品的吗？”海蛇问道。

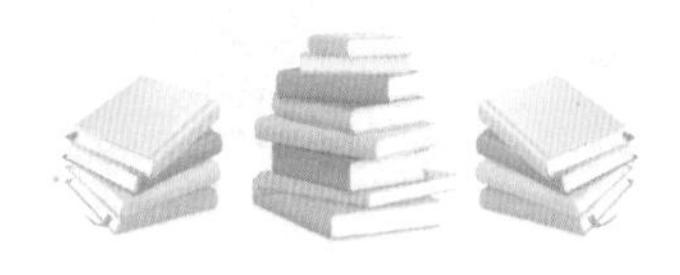

“Yes, I've seen the male making those sand sculptures before. Few people realise what great artists they are.”

“Flapping their fins, working tirelessly day and night, to create such perfectly geometric works of art!”

“Did you know that the young male creates his piece of art by crushing shells and using the pieces to decorate his sculpture?” Sea Snake asks.

完美的几何艺术作品！

Perfectly geometric works of art!

……碎贝壳还能提供至关重要的矿物质……

… crushed shells also deliver vital minerals …

“这些装饰品不仅漂亮，而且还能为河豚卵和幼鱼提供至关重要的矿物质。”

“真巧妙！在圆圈之间构筑的高埂能减弱水流。即使在无边无际的大海里，这些高埂也能防止小河豚被抛来抛去。”

“And not only are these decorations beautiful, but the crushed shells also deliver vital minerals to their eggs and young.”

“Ingenious! The little ridges built up throughout these circles, dampen the currents. Even though we are in the open sea, these help prevent the youngsters getting tossed around.”

“那么它肯定不仅仅是艺术了吧？我从未见过艺术与生活的基本需要结合得如此完美。”

“哦，是的。这是艺术与功能、美与实用性、心灵与头脑的结合，外加一剂热情和完全的奉献精神。否则河豚为什么要这么努力工作？”

“So surely this must be more than just art? I've never seen art combined in this way with a basic necessity of life.”

“Oh, it is. It is art and function, beauty and purpose, heart and brain, with a good dose of passion, and complete dedication. Otherwise why would one work so hard?”

……美与实用性、心灵与头脑……

… beauty and purpose,heart and brain …

为了打动自己心仪的女士……

Out to impress a lady ...

“也许河豚是为了打动自己心仪的女士……”虎鲨提示道。

“这或许是一个很好的理由。可是他在这里做雕塑的时候，还没有向任何异性表白过。”

“Maybe the pufferfish is out to impress a lady… ” Tiger Shark suggests.

“That may well be a good reason. But here he is, sculpting before he has even had a chance to propose to one.”

“我想这是一个关于真爱的故事，集艺术、工艺和对幸福健康家庭的渴望于一体。”

“可是谁会期望从一条既有毒又美味的鱼那里获得这份爱情呢？”

……这仅仅是开始！……

“I think it is a story of true love, combining art, craft and the desire to have a happy and healthy family.”

“Now who would have expected that from a poisonous delicacy?”

... AND IT HAS ONLY JUST BEGUN!...

……这仅仅是开始！……

… AND IT HAS ONLY JUST BEGUN! …

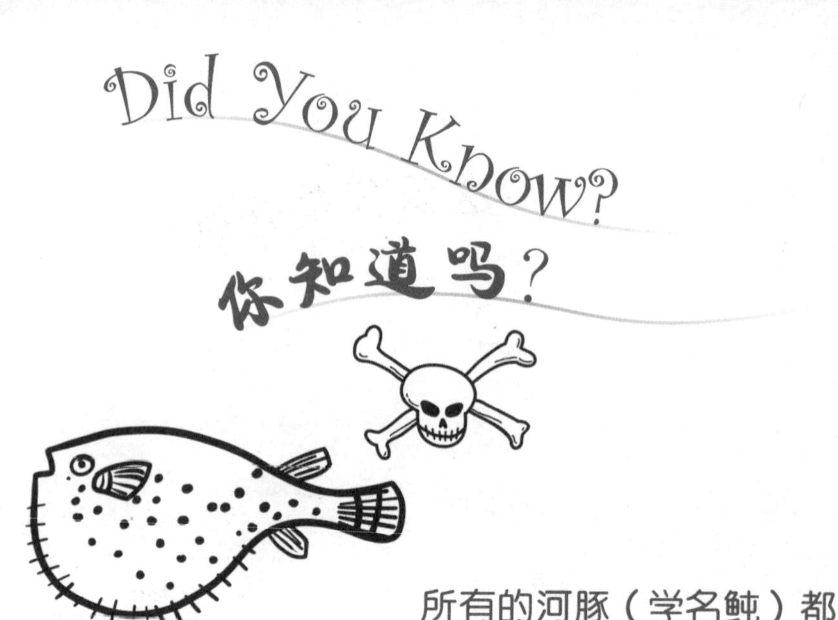

All pufferfish contain tetrodotoxin, a foul-tasting substance that can be deadly. It is up to 1,200 times more poisonous than cyanide. There's no known antidote.

所有的河豚（学名鲀）都含有河豚毒素——一种味道恶心的致命物质。它的毒性是氰化物的1 200倍。目前尚无解药。

Pufferfish meat is considered a delicacy. In Japan it is called fugu. It is expensive, and can only be prepared by trained and licensed chefs, who know that making an error would mean death. Many such deaths occur annually.

河豚肉被认为是美味佳肴。在日本，河豚被称为fugu。河豚很贵，并且只能由经过培训且有执照的厨师烹饪，一有闪失就意味着死亡。每年都有许多这类死亡案列。

There are more than 120 species of pufferfish worldwide. Most are found in tropical and subtropical ocean waters, but some species live in brackish and even fresh water.

世界上有超过120种河豚。它们大多生活在热带和亚热带海水中，但也有一些生活在半咸水甚至淡水中。

Large pufferfish feeding on clams, mussels, and shellfish, will crack these open with four fused teeth. These teeth keep on growing all their life. They trim their teeth by eating hard shells.

以蛤蜊、贻贝等贝类为食的大型河豚，会用 4 颗融合牙咬碎食物。这些牙齿一生都在生长。它们通过吃硬贝壳来修剪牙齿。

Young pufferfish wear a toxic 'cloak' that prompts predators to instantly spit them out. The female pufferfish passes the potent toxin to the developing embryos in her ovaries.

小河豚身上有一层有毒的"外衣"，促使捕食者立即将它们吐出来。雌河豚会将这种强毒素传给卵巢中正在发育的胚胎。

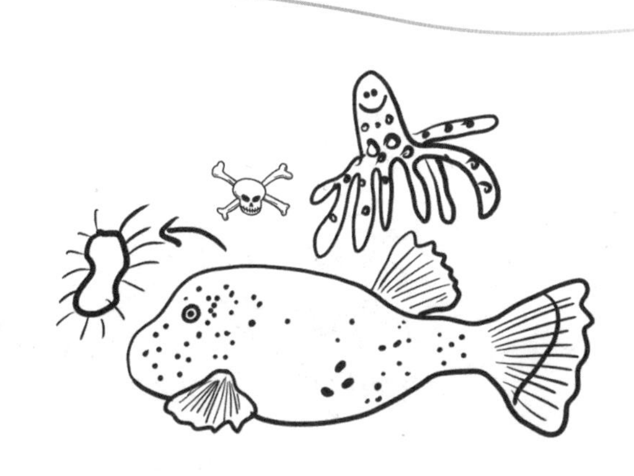

Poisonous pufferfish synthesise their deadly toxin from the bacteria in the animals they eat. These symbiotic bacteria are found not only in fugu but also in blue-ringed octopus, rough-skinned newts and some sea slugs.

毒河豚用它吃的动物体内的细菌合成致命的毒素。这些共生细菌不仅存在于河豚体内，也存在于蓝环章鱼、粗皮渍螈以及某些海蛞蝓体内。

Pufferfish do not have scales, but instead have spines that are very sensitive to changing water conditions. Pufferfish are susceptible to disease.

河豚没有鳞片，取而代之的是对变化的水环境非常敏感的刺。河豚易患病。

Japanese have eaten fugu for centuries. Fugu bones were found in settlements that date back 2,300 years. In China's Song Dynasty fugu was considered one of the 'three delicacies of the Yangtze'.

日本人吃河豚已经有几个世纪了。在 2 300 年前的定居点发现了河豚骨骼。在中国宋代，河豚被认为是“长江三鲜”之一。

Think about It

Would you make a piece of art to attract a life partner?

你会做一件艺术品来吸引你的人生伴侣吗?

Which would you choose: your stomach expanding to choke your predator, or being eaten and killing the one that eats you?

你会选择哪一种：把你的胃鼓起来使捕食者窒息，还是被吃掉并杀死吃你的捕食者?

Are you interested in pufferfish's work of art?

你对河豚的艺术作品感兴趣吗?

Fish do farm their own food, and are artists? Are you surprised?

鱼种植自己的食物，而且是艺术家? 你对此感到惊讶吗?

Make some drawings that depict the art discovered on the ocean floor. Ask friends and family members if they know how to make this in the sea. Then share information on the pufferfish, and engage them in a discussion: is this art, is this a way of life, is this an act of love, or just techniques to have the chance to live longer?

画几张描绘在海底发现的艺术品的图画。问问朋友和家人，他们是否知道这些艺术品是如何在海里创作的。然后分享关于河豚的信息，让他们参与讨论：这是艺术吗？这是一种生活方式吗？这是一种爱的行为吗？或者仅仅是一种延长生命的技术？

学科知识

Academic Knowledge

生物学	河豚的牙齿一直在生长；海蛇是一种有毒的蛇，不能在陆地上生活；虎鲨（学名鼬鲨）是夜行性捕食者；虎鲨有瞬膜；虎鲨是顶级掠食者；鲨鱼肝富含维生素A；人类的大脑在生物学层面上处理是否该冒险的问题；大脑的决策区由80%的兴奋细胞和20%的抑制细胞组成。
化　学	河豚毒素是神经毒素，它能阻断钠离子通道；河豚毒素用于治疗或缓解剧痛、心律失常、晚期癌症、偏头痛；信息素是一种传感物质。
物　理	河豚能自我膨胀是因为它的胃和腹部皮肤能够承受体积的大幅增加；海蛇的大脑对振动有听觉反应；虎鲨使用反荫蔽技术；便于急转弯的楔形头；侧线感觉器官可以感受微小振动。
工程学	棂条窗，一种精心制作的石质支撑物，就像里斯本大教堂展示的那样，它甚至能使玫瑰窗具有抗震能力；高埂以微小的干预阻止水流，效果却很大。
经济学	贝壳被认为是废物，但它能提供矿物质，因而成为培育肥沃土壤的关键；实用又漂亮的建筑具有特别的力量，还能带来可预测的结果；创业投资是没有确定结果的投资。
伦理学	人们捕杀虎鲨是为了获得它们的鳍，其他部位则被丢弃；选择性捕杀对减少人类和虎鲨之间的遭遇不起作用，同时鲨鱼数量仍大幅减少；认为只有人类才能创造艺术的观点。
历　史	在日本的古代垃圾堆中发现的河豚骨，至少可以追溯到2 800年前；玫瑰窗被发现在罗马时代就存在了，比如万神殿的圆顶（建于公元125年），并成为哥特式建筑的特征。
地　理	海蛇生活在印度洋和太平洋的沿海水域；太平洋中部岛屿上的虎鲨。
数　学	玫瑰窗图案借鉴了罗盘和直尺；围绕一个圆环等距离分布的12个点，构成一个十二角星。
生活方式	考虑到每天有成千上万的人游泳、冲浪、潜水，鲨鱼攻击人类的概率很低；生活品质，艺术与功能，美观与实用，心与脑。
社会学	多种生存策略，比如河豚利用毒液和充气。
心理学	无所畏惧的探索，比如跳伞，或在鲨鱼身边潜水。
系统论	虎鲨以垃圾为食，吞食人类留在海水中的废弃物。

情感智慧

Emotional Intelligence

海　蛇

海蛇向虎鲨发起挑战，并且调侃虎鲨。他很勇敢，因为虎鲨比他大得多。海蛇以一种轻松的方式接近虎鲨，虎鲨也容忍他个性鲜明而且自信的姿态。海蛇把话题转向河豚的危险性，以巧妙方式试探对方，促使虎鲨不吃河豚。但这也不奏效，因为虎鲨不受河豚毒素影响，海蛇遂将话题转到艺术上。他阐述河豚的艺术能力，希望唤起虎鲨对河豚的同情，促使虎鲨不吃河豚。这一策略似乎奏效了。随后他们又探讨了危险和艺术的话题，使对话更有深度。海蛇强调心（激情）和脑（推理）的结合。然而，最后海蛇再次确认了他认为河豚是一种美食。

虎　鲨

虎鲨非常自信，也很务实，意识到自己不是唯一喜欢吃河豚的物种——人类也视河豚为美味佳肴。她对河豚对人类的致命毒性感到惊讶。她观察敏锐，并且非常欣赏河豚的艺术作品。她认为河豚很勤劳，并对河豚这种勤劳的生物产生了同理心。她将美与实用性联系起来。虎鲨开始尝试理解河豚想要打动自己意中人的愿望，并且不再想吃掉河豚。

艺术

The Arts

让我们学习如何用沙子创作艺术品。这是在物体表面上倾泻沙子的艺术，你可以用原色沙子，或者用其他颜色。如果沙子没能固定住，你可以重新绘制，或者局部改绘。用摄像机来记录整个创作过程，这样别人就可以观察这种独特艺术的创作过程了。如果你想要真正的挑战，试着用湿的沙子。你很快就会更加赏识河豚的技能了。

思维拓展

Systems: Making the Connections

河豚有独特的防御技术，使它有别于其他鱼类。它能够用水和空气填充身体使自己鼓胀，使身体变得更大。它没有鳞片，但有长而坚硬的刺，这种防御机制会让任何捕食者窒息，迫使捕食者吐出这种膨胀了的多刺鱼。除了独特的防御机制，河豚现在也被认为是海洋中最伟大的艺术家之一。雄河豚创作的沙圈表明了他对配偶和后代的奉献和承诺。由于行动能力有限，河豚游得很慢，雄性河豚会用小鳍在海里画出最壮观的“沙画”来吸引配偶，并为后代创造一个适合生存的栖息地。所有河豚都使用相同的核心技术和必备技能，最终作品却总是不同。雌河豚很欣赏这种艺术，她选择伴侣的标准是雄河豚所构筑的巢穴的美观性以及保障后代安全的能力。所以，这不只是艺术表达，还有重要实用目的。好的艺术家不仅让我们敬畏他们卓越的技艺，而且还能通过语言之外的方式诠释我们生活的时代。河豚绘制的细小线条和圆圈不仅是一种令人惊叹的表现形式，而且在确保鱼卵和刚孵化的幼鱼免受水流冲击方面起着重要作用。河豚知道为后代创造安全空间所需要的精确高度和坡度。雄河豚会用它的4颗坚固的融合牙碾碎贝壳，并把贝壳碎片嵌在它的巢穴的高埂上。这些贝壳碎片创造了一个局部的高碱性环境，就好像河豚在保护它的幼鱼免受海洋酸化的威胁。科学家注意到，只要有碎贝壳，微量营养物就会在这种富含矿物质和碳酸钙的环境中生长。河豚似乎为下一代创造了理想的环境，为它们提供保护和营养。所以，雄河豚不仅是艺术家，也是农夫，是栖息地的建设者，是尽职尽责的父亲。谈起河豚，我们的重点不应该放在河豚毒素的杀伤力上，我们也应该了解这种行动缓慢而笨拙的小鱼是如何成功地处理日常事务，吸引配偶，并照顾好他们的后代的。

动手能力

Capacity to Implement

我们也需要保护弱者和幼小者，并确保粮食安全。因此，我们持续研究海浪，寻求将损害降到最低的办法。我们能从河豚身上获得什么启发？想想河豚建造的那些隆起的高埂，这些高埂会对海浪产生什么影响？看看你能想出什么主意来减弱海浪的力量。